# ULTIMATE PREPPER'S SURVIVAL BIBLE

*[30 In 1] The New Health, Safety and Environmental (HSE) Worst-Case Scenario Survival Guide, Life-Saving Strategies to Be Self Sufficient and to Keep Your Family, Workplace and Factory Safe in Every Emergency*

*By Jackson W. Everhart*

Copyright © 2024 by Jackson W. Everhart

All rights reserved. No part of this publication may be reproduced, distributed, or transmitted in any form or by any means, including photocopying, recording, or other electronic or mechanical methods, without the prior written permission of the publisher, except in the case of brief quotations embodied in critical reviews and certain other noncommercial uses permitted by copyright law. For permission requests, write to the publisher, addressed "Attention: Permissions Coordinator," at the address below.

**Publisher:** Everhart Publishing

244 Survivalist Lane

Prepper City, USA 56789

**Ordering Information:**

Quantity sales. Special discounts are available on quantity

purchases by corporations, associations, and others. For details, contact the "Special Sales Department" at the address above.

First Edition, 2002

Library of Congress Cataloging-in-**Publication Data:**

Everhart, Jackson W.

The Ultimate Prepper's Survival Bible: The Complete Health, Safety, and Environmental (HSE) Worst Case Scenario Survival Guide, Life-Saving Strategies to Be Self-Sufficient and Keep Your Family, Workplace, and Factory Safe in Every Emergency / Jackson W. Everhart.

p. cm.

Includes bibliographical references and index.

1. Survival skills. 2. Emergency management. 3. Health and safety. 4. Self-sufficiency. I. Title.

HV551.2. E94 2024

613.69--dc23

Cover design by Fred's printings

Interior design by Collins' Designs

This book is dedicated to all the brave individuals who strive to be prepared for any eventuality and ensure the safety and well-being of their loved ones and communities.

## Table Of Content

*Chapter 1 HSE Level 1 (HSE Appreciation)* ...................................................8

Introduction

Definition

Scope and Application

Group standards and Requirements

Unsafe Acts and Unsafe Conditions

Management system

HSE Responsibilities and Competence

*Chapter 2 HSE Level 2 (General HSE)* ..............................................................56

Job Safety Analysis

Personal Protective Equipment

Procurement and Assurance of PPE

Introduction to Security

Security Survey and Risk Assessment

Security Plan

Permit to Work

Complementary Certificate

Fire Prevention, Protection and Evacuation Procedure

Classes of Fire

Starvation

Identification of Fire Extinguishers, Size and Color codes

*Chapter 3 HSE Level 3 (Supervisory HSE)..................................................................85*

HSE Management System:

Introduction to HSEMS

HSE Commitment

Elements of HSEMS

Hazard and Effect Management Process (HEMP)

Principles of HEMP

Risk Management

Risk Assessment Matrix

Structured Review Techniques

First Aid

Cardiopulmonary Resuscitation (CPR) Practical

Environmental Management

Environmental Management in Oil Company

Environmental Pollution

Air Pollution

Water Pollution

Soil Pollution

Waste Management

Environmental Management

*About the Author............................................174*

# CHAPTER 1 HSE LEVEL 1 (HSE APPRECIATION)

## Introduction

### WHY PREPARE FOR THE WORST?

In an increasingly unpredictable world, the ability to be self-sufficient and prepared for any emergency is not just a luxury but a necessity. "The Ultimate Prepper's Survival Bible: [30 in 1] The New Health, Safety, and Environmental Worst-Case Scenario Survival Guide" is designed to be your comprehensive resource for surviving and thriving in the face of disasters.

Whether you're at home with your family, in the workplace, or managing a factory, this guide offers life-saving strategies to ensure safety and self-sufficiency.

## THE IMPORTANCE OF PREPAREDNESS

We live in a time where natural disasters, health crises, and environmental hazards can strike without warning. From hurricanes and earthquakes to industrial accidents and pandemics, the threats we face are diverse and potentially devastating. The recent global events have underscored the need for individuals and communities to be better prepared. By adopting a proactive approach, we can mitigate the impact of these emergencies and protect our loved ones.

Preparedness isn't about living in fear; it's about empowering yourself with the knowledge and tools needed to handle adverse situations effectively. This book aims to shift your mindset from reactive to proactive, providing you with the confidence and capability to face any challenge head-on.

## What You Will Learn

This guide is divided into carefully structured chapters, each focusing on a critical aspect of emergency preparedness:

*Understanding Health, Safety, and Environmental (HSE) Principles:* Learn the foundational concepts of HSE and why they are crucial in everyday life. This chapter covers risk assessment, management, and the importance of maintaining safety standards.

*Building a Comprehensive Emergency Plan:* Discover how to create tailored emergency plans for your family, workplace, and factory. Effective communication, evacuation routes, and regular drills are emphasized to ensure everyone knows what to do in an emergency.

*Essential Survival Skills:* Equip yourself with vital skills such as first aid, food and water storage, and building temporary shelters.

These skills are the bedrock of survival, helping you to stay safe and healthy during crises.

*Environmental Preparedness:* Understand how to prepare for natural and man-made disasters. From earthquakes to chemical spills, this chapter provides detailed strategies to minimize risk and recover swiftly.

*Self-Sufficiency Strategies:* Learn how to become self-reliant through sustainable practices such as growing your own food, harnessing alternative energy sources, and purifying water. These strategies not only enhance your survival prospects but also promote long-term sustainability.

*Communication and Information:* Effective communication is critical during emergencies. This chapter covers how to stay informed, use technology to your advantage, and build a supportive community network.

*Psychological Preparedness:* Addressing the mental and emotional aspects of emergencies, this chapter offers techniques for managing stress, maintaining mental health, and supporting your family and community through tough times.

*Ongoing Preparedness and Improvement:* Preparedness is an ongoing journey. This final chapter emphasizes the importance of continuously updating your plans, learning from experiences, and encouraging a culture of preparedness in your community.

## The Structure of the Book

Each chapter is designed to be informative, practical, and easy to understand. The content is enriched with real-life examples, detailed explanations, and actionable steps. Additionally, the book includes appendices with quick reference guides, checklists, and a resource directory for further reading. A glossary of key terms and a comprehensive index make it easy to navigate and find specific information.

## What You Need To Know About HSE

### What you need to know about HSE.

Health, Safety and Environment (HSE) is a general term for regulations, methods and processes designed to help protect workers, the environment and the public from harm. The methods and processes in HSE are learned through training and re-training of those exposed to hazards.

A safe work environment without hazards and accidents can be achieved through effective implementation of a company-wide health and safety (HSE) management system. Thus, training and re-training personnel on the global best practice in health and safety cannot be ignored.

### Why should I register for she training?

You should definitely get trained and certified in she as a basic requirement by law for employers of labor. You automatically become an asset to your organization, and it brightens your prospects.

## Definition Of Terms

### What is HARM

It is Physical, mental or moral impairment or deterioration. It includes death, injury, physical mental ill health, damage to property, loss of production or any combination of these.

### What Is HAZARD

A source or a situation with a potential to cause harm, including human injury or ill health, damage to property, damage to the environment, or a combination of these. A hazard is an act of conditioning in the workplace that could result in an injury or loss if it remains uncorrected.

### INCIDENT

An unplanned, undesirable event that adversely affects completion of a task.

## ACCIDENT

An unplanned or undesired event that can result in personal injury, property damage or harm the environment

## What is NEAR MISS

A near miss is an event that has occurred where there has been no loss or injury but under slightly different circumstances an event with consequences could have occurred. Incidents were no property was damaged and no personal injury sustained, but where, given a slight shift in time or position, damage and/ or injury easily could have occurred, or an event where no contact or exchange of energy occurred and thus did not result in personal injury, asset loss or damage to the environment.

A near miss is an event has occurred where there has been no loss or injury but under slightly different circumstances an event with consequences could have occurred.

## DANGEROUS OCCURRENCE

Escape of flammable substance, explosion, fire, and pipeline ruptures, transport incidents, etc.

## EXPOSURE

The measurement of time during which the subject is at risk from a hazard.

## FAT

Factory Acceptance Testing (FAT) is a critical process used in Health, Safety, and Environment (HSE) management, particularly in industrial and manufacturing settings. It involves the testing and verification of equipment and systems before they are shipped to the site where they will be installed and used. The goal of FAT is to ensure that the equipment meets all specified requirements and functions correctly in a controlled factory environment before deployment.

**Factory Acceptance Testing**

## *FATALITY*

Death due to a work related incident or illness regardless of the time between injuries or illness and death.

## *HOUSEKEEPING*

Maintaining the working environment in a tidy manner.

## *INCIDENT POTENTIAL*

Incident potential refers to the likelihood or probability that an event or situation could lead to an incident or accident. It's about assessing how likely something is to cause harm or disruption.

Understanding incident potential involves recognizing hazards and evaluating how likely they are to result in an actual incident if not addressed.

It's a way of identifying risks and taking preventive measures to reduce the chances of accidents or injuries occurring.

## INTERFACE DOCUMENT

A document that clearly identifies how the Owners HSE expectations and the Shipyard's HSE management system will be interlinked during the work programmed.

## LOST TIME INJURY (LTI)

Accident resulting in time off work. It is a work related injury or illness that renders the injured person unable to perform any of their duties or return to work on a scheduled work shift, on any day immediately following the day of the accident

**Personal Protective Equipment (Ppe)**

All equipment and clothing intended to be utilizes, which affords protection against one or more risks to health and safety. This includes protection against adverse weather condition

## RESTRICTED WORK CASE

work related injury or illness that renders the injured person unable to perform all normally assigned work functions during a scheduled work shift or being assigned to another job on a temporary or permanent basis on the day following the injury

## RISK

A measure of the likelihood that the harm from a particular hazard will occur, taking into account the possible severity of the harm.

## RISK ASSESSMENT

The process of analyzing the level of risk considering those in danger, and evaluating whether hazards are adequately controlled, taking into account any measures already in place.

## RISK MANAGEMENT

The process of identifying hazards, assessing risk, taking action to eliminate or reduce risk, and monitoring and reviewing results.

## TRAINING

The process of imparting specific skills and understanding to undertake defined tasks.

## UNSAFE ACT CONDITION

Any act or condition that deviates from a generally recognized safe way or specified method of doing a job and increases the potential for an accident.

## SAFETY ORIENTATION

Briefing on facility and its processes

## SITE SPECIFIC ORIENTATION

Briefing on the hazards present in a particular job

## DOSIMETER

A gas detector is a small instrument, such as a film badge, used to indicate and measure amounts of radiation.

## OXYGEN ANALYZER

An oxygen analyzer is a small hand – held instrument used to measure the percentage of oxygen in air.

## AUTOMATED EXTERNAL DEFIBRILLATORS (AEDs)

Automated external defibrillators (AEDs) are small devices that will send electrical pulses through wires attached to chest pads to cause an electrical shock to a person's heart to make it pump on its own.

## SWL

Safe Working Load.

## WORK PROGRAMMED

The work being undertaken by a site on behalf of the company.

## WORKSITE

The premises where any building operations or works of engineering construction related to the work program are being carried out.

## Incident Potential Severity Rating (Ipsr)

A rating of the damage or injury potential of an incident based on the severity of the worst credible outcome, faking all circumstances of an incident (including the control and consequence mitigating measures in place) into consideration. The risk assessment determines the composition of the investigation team and review panel.

### INCIDENT NOTIFICATION FORM

This is a standard form to be completed on the web or alternatively on paper if e-mail is not available to the reporter, by the affected line supervisor as a preliminary report to provide the salient features of the incident, including its potential severity rating.

It should be forwarded within 24 hours to be appropriate HSE Department and up the lines as required.

*PERSONAL PROTECTIVE EQUIPMENT (PPE)*

What is PPE

These are protective gear worn by workers to minimize impact or prevent injuries in the event of an incident while carrying out activities

It is essential to note that.

1. Foot guards must be worn when using jacks, hammers, tampers, and similar
2. Equipment safety belts (or harnesses), lifelines, and lanyards should be worn while working elevation is 3m from ground or platform level.
3. Any lifeline, safety belt, or lanyard actually subjected to in − service loading as distinguished from static load testing shall be

immediately removed from service and shall not be used again for workers' safe guarding.
4. Lifelines shall be secured above the point of operation to an anchorage or structural member.
5. Safety belt lanyard shall be a minimum of 14mm nylon, or equivalent with a proper length of falling distance no greater than 1.8m

## Hemp (Hazard And Effect Management Process)

### *DEFINITION:*

A hazard is the potential for harm. In practical terms, a hazard often is associated with a condition or activity that, if left uncontrolled can result in an injury or illness. Hazards can be identified and assessed through brainstorming (what could happen), reviewing past experience and deviations from standards.

### *RISK*

Risk is defined as a function of the probability of occurrence of an undesired event together with a measure of its advice consequences. It is the exposure to a hazard. The level of risk is determined by:

- The severity of the consequences, and
    - The probability of the incident occurring
- Risk = Probability x Severity

*Consequences of the outcome of an action or deviation from works standards*

1. Personal injuries or ill health effects
2. Spills or environmental damage
3. Public disruption or adverse media
4. Financial loss fires/explosion, equipment damage, operating disruptions, etc.
5. Cost of emergency repairs
6. Loss of a unit or plant revenue stream
7. Cost of meeting contract obligations with other units or companies

9. Severity is the size of consequences
    a. the actual or potential degree of harm or loss
    b. Fatality, disabling injury, medical treatment, first aid, etc.
    c. release size, equipment damaged

*Probability Is the likelihood an incident will occur within a given time frame*

Normally shown as a value between 0.0 (%) and 1.0 (100%)

Often described as average length of time for the incident to occur

**Risk Analysis**

This is the development of a qualitative estimate of risk or how much risk is involved in performing a task.

## *Risk assessment*

Evaluation of risk analysis results to support risk decisions, through comparison with risk criteria or other standards of acceptability

Is this level of risk acceptable? or

Does this have more/less risk than alternatives?

## *Risk Management*

Process when identifying, assessing and controlling risk to protect human life, the environment, physical assets and company reputation in a cost- effective manner.

*Reasons why risk are assessed:*

*Risk is assessed to:*

1. Compare alternatives, particularly in the design phase
2. Optimize the use of scarce resources, money, people and time
3. Identify hazardous situations /procedures
4. Provide knowledge of patterns of events and identify critical parts of the operation.
5. To provide information required to make decisions on whether the level of risk is tolerable
6. Meet regulatory requirements
7. Provide a clear framework in which all available information is used
8. Identify potential economic vulnerability
9. Risk management deficiencies?
    a. Inadequate job planning

b. Inadequate communication/handover (Management of change)
c. Inadequate permit – to – work system
d. Failure to assess all risks inherent in a job task
e. Inadequate emergency response training/ planning
f. Production pressure

10. Risk Assessment & Management Responsibilities:
   a. Every worker has a responsibility to:
   b. Identify Hazards
   c. Assess the risks from the hazards
   d. Manage the risks
   e. Follow up / verify solutions

*Typical Methods for Identifying and Assessing Base Business Hazards:*

1. Brainstorming (what could happen?)
2. Reviewing past experiences – incidents, near misses, hazard hunts (what has happened?) Deviations from standards (task observations, site walk – through)
3. JSAs, inspections, hazard hunts
4. Routine/ Temporary changes (MOC)
5. More formal risk assessment if complex issues are not easily resolved
6. Steps in hazard and effect management process (HEMP)

*Identification of hazards:*

Are people, environment or asset expose to potential harm?

To identify and assess hazards, employers and workers:

1. Collect and review information about the hazards present or likely to be present in the workplace.

2. Conduct initial and periodic workplace inspections of the workplace to identify new or recurring hazards.

3. Investigate injuries, illnesses, incidents, and close calls/near misses to determine the underlying hazards, their causes, and safety and health program shortcomings.

4. Group similar incidents and identify trends in injuries, illnesses, and hazards reported.

5. Consider hazards associated with emergency or no routine situations.

6. Determine the severity and likelihood of incidents that could result for each hazard identified, and use this information to prioritize corrective actions.

As we go about our daily tasks, it's important to remember that safety should always be our top priority. Accidents can happen due to unsafe acts or unsafe conditions, which can lead to injuries, property damage, and even loss of life. In this toolbox talk, we will discuss what constitutes an unsafe act and an unsafe condition, and provide examples to help you identify and avoid them.

One of the "root causes" of workplace injuries, illnesses, and incidents is the failure to identify or recognize hazards that are present, or that could have been anticipated. A critical element of any effective safety and health program is a proactive, ongoing process to identify and assess such hazards.

Unsafe conditions are any factors in the workplace that increase the risk of accidents or injuries. These could include poor lighting, cluttered workspaces, slippery floors, or faulty equipment.

Preventing unsafe acts and unsafe conditions is everyone's responsibility. Here are some tips to help you prevent accidents on the job:

Always use proper tools and equipment for the task at hand

- ✓ Wear proper PPE as required by your job
- ✓ Follow established safety procedures and guidelines
- ✓ Keep your workspace clean and organized
- ✓ Report any unsafe conditions or hazards to your supervisor
- ✓ Never take shortcuts or risks when performing a task
- ✓ Take breaks as needed to stay alert and focused

## Unsafe Act Auditing

Time after time, accidents occur because the established safe procedures and practices have been ignored or forgotten. Investigation will often reveal that the basic cause of an accident has been the steady deterioration of working practices since a plant was installed.

It is incredible how quickly a well-constructed plant with clear procedures and safeguards can become unsafe without management or workers being aware. A short cut in procedures becomes accepted practice; a small malfunction of equipment is overcome by the operator and the 'new' procedure adopted as standard. And so it goes on, until the inevitable accident occurs and the manager cannot understand how he came to walk past the problem every day without seeing the hazard. It is because of this that all responsible management sees the need to promote systematic and regular auditing of their safety procedures and practices.

It is the overall responsibility of management to ensure that operational related activities are adequately controlled. The responsibility covers both the human and physical elements, as accidents are often caused by failure to achieve control in these areas; accidents are not, as is often believed, the result of straightforward failures of technology, but also result from social, organizational and technical problems.

A major part of this responsibility is to make sure that personals will not be exposed to hazards which they are unaware of or are unable to control. This implies high technical competence, but on its own it will not ensure a consistently safe place of work.

The behavior of the human being in an advanced technological working place is also of fundamental importance.

Hazards may cause accidents. One definition of hazards is that they are unsafe art of persons, unsafe mechanical conditions or unsafe environmental conditions. An unsafe acts audit is a scheduled inspection of a plant area, conducted by the plant supervisor or plant management member, one of the plant employees and a company safety officer.

*Unsafe Act Audit May Cover the Following Subject:*

- ✓ House keeping
- ✓ Unsafe acts of conditions

It will use lists referring to specialized equipment and procedure according to activity. The result of the safety inspection, provided by the checklist filled in by the visiting team, will be handed over to the plant manager who will be responsible for monitoring all resultant actions and follow-up.

Unsafe act audit is aimed at the reduction of exposure to all of the aforementioned conditions as part of an overall safety programmer.

*Reasons for Unsafe Act Auditing:*

- Unsafe act audits are intended to promote, improve and then maintain good safety performance.
- Unsafe act audits (unlike accident statistics, which only record past accidents) are an instrument for the direct prevention of accidents, as immediately generate position action across the whole business activity.
- Unsafe act audits are a position means by which performance can be further improved. They are a structured and effective (review of the total factory which emphasizes the need for good housekeeping and planned maintenance) way of preventing accidents and controlling safety hazards.

## *Objectives of safety Audits:*

The objective of an unsafe act audit is to evaluate the effectiveness of company's safety effort and make recommendations which read to a reduction in accidents and minimization of loss potential. Unsafe act audit is carried out in any work environment basically to:

- Increases Safety Awareness
- Raise Standards, Reaffirm and Improve the Accepted Standards of HSE Performance
- Promote Communication and Understanding
- Motivate Employees Subordinates
- To Change the Cultural Attitude Towards Safety

## Allocating Resources For Unsafe Act Auditing

Regular audits should be based on the premise that resources should be made available to identify and eliminate hazards before accidents occur, rather than use the resources of manpower and materials only after injuries and damage to equipment have resulted in human suffering significant monetary loss which, in certain circumstances, affects the profitability severely. The safety audits will monitor all activities performed on site, and in particular:

- The basic safety policy and organization of the company.
- Management commitment and examples on safety matters.
- Administration and safety activity.
- Accident reporting and investigation
- Opportunity of injury and record of every injury.

- Safety committees.
- Working rules and practices for each company location, including visitors and contractors.
- Compliance with statutory regulations and company standards.
- Behavior and unsafe acts of personnel and their relationship to compliance with safety rules.
- Activity related certification of employees.
- First aid certified employees.
- Training needs and activities.
- Hazard review of process equipment for either new or existing facilities.
- Operating procedures.
- Safety work permits.
- Emergency procedures.

## SAFETY PRINCIPLES/FACTS

- Almost 98% of accidents are a result of unsafe acts and conditions
- If it is not safe, it is not right
- People are the most critical element in the success of a safety program
- Accident prevention is loss prevention (cost of repairs and treatment)
- No job is so important that it can be done at the expense of safety
- While risk exists, accidents and injuries are preventable
- Safety is every body's business
- Management is responsible for preventing/controlling injuries and occupational illness
- All work pace accidents should be investigated, and lessons shared
- Training is an essential element for safe workplace practice

### Technique For Conducting Unsafe Act Auditing:

1. *PREPARATION*
2. *OBSERVATION*
3. *DISCUSSION*
4. *RECORDING*
5. *FOLLOW UP*

## PREPARATION:

In view of reaching the aims described, every unsafe act audit should be carefully prepared. Questionnaires should be established for each activity and plant by the audit team and be approved by company management. Unsafe act audit questionnaires will also include questions on general organization, management and training.

Previous audit reports must be studied prior to the next audit to ensure that all actions from the previous audit either have been carried out satisfactorily or are included as an additional item on the current audit report. Audits should be carried out frequently and regularly, aiming for a total of one to two hours per week per person. The time should be devoted solely to safety inspections and not combined with other activities.

## OBSERVATION:

Before carrying out an unsafe act audit, stop, observe the work environment and work activity in progress. Decide on what to do, act on it and report. Concentrate on the action of people working and not on things. Develop a questioning attitude to determine what injuries might occur if the unexpected happens and how it will affect the job or better ways of doing the job.

## DISCUSSION:

Interact with workers and find out how the job could be done more safely. Make them know that the audit is for improvement and not appraisal so they can willingly identify work hazards and the way forward. Ensure that the unsafe acts and conditions observed are discussed, and good safety practices commanded.

## RECORDING:

Unsafe acts observed should be listed, discussed and corrected. Items for further actions should be noted and copies made available for follow up. Local management must be involved in the review of the findings of the audits team before the audit team leaves site.

*Following the audits, a report should be issued containing the following.*

1. A description of all findings relative to items needing a proposal for remedial action

2. A description of any defects detected on equipment and initial proposals for remedial actions.

3. The names and job titles of those people who are responsible for initiating remedial actions.

4. The names and job titles of those people who are responsible for initiating remedial actions.

5. The need for any revision of operating instructions or company standards.

6. Agreed target dates for completion (which realistically allow sufficient time for thorough technical assessment and consequent changes, if necessary).

7. Copies of the audit report shall be given to the local plant manager and to the company management, who will take decisions and supervise follow-up actions agreed by the safety audit team.

## FOLLOW UP:

Monitoring of approved safety audit conclusions and recommendations is an important activity to ensure improvement of the safety level of a plant or company. It is the responsibility of the plant manager to see that audit conclusions and recommendations are implemented by the agreed target dates.

## UNSFE ACT AUDIT TEAM:

Unsafe act audit should be carried out at general management level and at each plant or site. The audit team members should be carefully selected for their knowledge and experience in the field of audit, from general management, plant management, and other safety specialists. For example, at plant level, the team should consist of, as a minimum:

- *Site plant manager*
- *Site plant foreman*
- *Safety specialist*

As unsafe act audits are carried out at a number of levels in a factory, E.g. small department, followed by operating and then general management level, it is essential that a team member at the lowest level is also incorporated in the next level team, and so on, so as to ensure a common approach, improved ease of reporting and communication.

## QUALITIES OF A GOOD OBSERVER:

- Is safety conscious
- Record observation systematically
- Not satisfied with over view impressions
- Is selective
- Know what to look for
- Asking: why? what if?
- Keep an open mind
- Look deeper into things

# CHAPTER 2 HSE LEVEL 2 (GENERAL HSE)

## Basic First Aid

When accidents happen that requires serious medical help, more times than not, basic care will need to be given before advanced medical help can arrive.

First aid is basic medical care given by a person until advanced medical help can arrive. Basic medical care includes techniques such as performing rescue breathing, performing CPR, helping a choking victim, bandaging wounds, splinting limbs, and treating shock.

It is the skilled application of accepted principle of treatment on the occurrence of any injury or sudden illness, using facilities or materials available to you at any time. It is the approved method of treating a casualty until places, if necessary in the care of a doctor or removed to hospital.

A first aid program will teach how to handle emergency situations, how to protect you while giving first aid, and how to care for many types of illnesses and wounds. You will also learn techniques that can save lives and procedures to follow after performing first aid.

Besides wearing appropriate first aid PPE, which will be discussed later in this chapter, first aid givers and other responders who will be entering a contaminated or hazardous area must be briefed on the hazards they may encounter and ways to protect themselves from those hazards before entering the area.

**Emergencies That Present And Immediate Threat To Life Are**

1. Obstruction in the airway
2. Cessation of breathing (Respiratory Arrest)
3. Cessation of heart (Cardiac Arrest)
4. Emergency medical evacuation procedure

## OBJECTIVE OF FIRST AID

First aid program is essential in any work environment to:

1. Save or preserve life.
2. Maintain an open air way by positioning the casualty correctly.
3. Begin artificial Respiration, if the casualty is not breathing and heart is not beating, and continue treatment until skilled medical aid is available.

4. Control breathing
5. Prevent the condition worsening, that is, to prevent further injuries and complication

## FIRST AID KITS

1. Adhesive tape

2. 4" x 4" sterile gauze pads

3. Antacid - for indigestion

4. Antidiarrheal (Imodium, Pepto-Bismol, for example)

5. Antihistamine cream

6. Antiseptic agent (small bottle liquid soap) - for cleaning wounds and hands

7. Aspirin - for mild pain, heart attack

8. Adhesive bandages (all sizes)

9. Diphenhydramine (Benadryl) - oral antihistamine

10. Book on first aid

11. Cigarette lighter - to sterilize instruments and to be able to start a fire in the wilderness (to keep warm and to make smoke to signal for help, for examples)

12. Cough medication

13. Dental kit - for broken teeth, loss of crown or filling

14. Exam gloves

15. Small flashlight

16. Ibuprofen (Advil is one brand name); another good choice is Naprosyn (Aleve is a brand name)

17. Insect repellant

18. Knife (small Swiss Army-type)

19. Moleskin - to apply to blisters or hot spots

20. Nasal spray decongestant - for nasal congestion from colds or allergies

21. No adhesive wound pads (Telfer)

22. Polypore antibiotic ointment

23. Oral decongestant

24. Personal medications (enough for the trip duration and perhaps a couple of extra in case of delays) and items (for example, a cane or knee braces if needed)

25. Phone card with at least 60 minutes of time (and not a close expiration date) plus at least 10 quarters for pay phones and a list of important people to reach in an emergency; cell phone with charger (cell service is not available in many areas, especially remote areas)

26. Plastic resalable bags (oven and sandwich)

27. Pocket mask for CPR (although now, CPR does not have to be mouth to mouth)

28. Safety pins (large and small)

29. Scissors

30. Sunscreen

31. Thermometer

32 Tweezers

33 A list of yours and other family member's medical history, medications, doctors, insurance company, and emergency contact persons (this can be accomplished easily with a flash drive)

## Occupational Health

Health is a state of wellbeing with the absence of illness or disease.

According to World Health Organization (WHO), Occupational Health is the promotion and maintenance of the highest degree of physical, mental and social well –being of workers in all occupations by preventing departures from health, controlling risks and the adaptation of work to people, and people to their jobs. It is the relationship between work effect on health and vice versa.

Occupational health helps employers care for and understand the need of their employees, creating an enabling business environment and as well optimize staff performance and productivity.

Occupational health programs cut through, hearing conservation, substance abuse testing and control, health hazard awareness, fitness for duty assessment, etc.

## Occupational Health Hazards

These are acts and conditions that reduce productivity and efficiency in the work place. They include:

- Physical Hazards
- Noise
- Temperature (heat/cold)
- Poor illumination
- Vibration
- Chemical
- Liquid (e.g. acids)
- Poisonous gases (CO- Carbon monoxide toxic fumes)
- Harmful dusts
- Biological
- Infectious diseases such as cancer, dermatitis etc.
- Ergonomic

- Badly designed work area, tasks and equipment.
- Mechanical
- Manual handling, being hit by a moving equipment etc.
- Psycho-social
- Lack of job satisfaction,
- Insecurity
- Poor interpersonal relations
- Work pressure

## Noise Hazard And Hearing Conservation

Prolonged, unprotected exposure to high levels of noise can result in permanent hearing loss. Prolonged high noise exposure causes deterioration and loss of cilia hair cells and their supporting structures. Hearing loss occurs *g-r-a-d-u-a-l-l-y* with no pain or warning signs.

### Sound and Noise

Sound results from air pressure waves generated by vibration or movement that we detect with hearing mechanisms while noise is Unwanted Sound Frequency. Weighted sound pressure is measured in units of decibels (dB).

## *EFFECTS OF HIGH NOISE*

1. Causes stress and anxiety

2. Interferes with speech and ability to communicate

3. May cause sleep difficulties

4. Temporary hearing shifts

5. Pain (very high levels)

6. Can damage hearing

## *HEARING CONSERVATION*

When an employee gets hired, an audiometric test is done to get a base line of his/her hearing. Then tests are conducted on time basis to identify if the employee is beginning to be affected by exposure to excessive noise.

Wear hearing protection in high noise areas, when using high noise high noise power tools, when working near temporary high noise jobs areas or sources.

For very high noise levels; 95 dB and higher, use double hearing protection.

## Ergonomics

US department of Health defines Ergonomics as the science of fitting workplace conditions and job demands to the capabilities of employees.

It is an applied science concerned with designing and arranging things people use so that the people and things interact most efficiently and safely. In an office environment, typically it is adapting your computer workstation to fit your individual needs.

Ergonomic principles are used to improve the "fit" between the worker and the workplace.

A practical approach to Ergonomics considers the match between the people, the equipment; they use the work processes and the work environment. Person's capabilities, physical attributes and work habits must be recognizing to improve ergonomic factors in the workplace

## ERGONOMIC RELATED INJURIES

When Ergonomics are improved in the workplace, we work smarter, not harder. Ergonomic related injuries include:

- CTD's (cumulative trauma disorders)
- RSI's (repetitive stress injuries)
- RMI's (repetitive motion injuries)
- Which are all considered:
- MSD's (musculoskeletal disorders)
- MSD's can affect muscles, tendons, nerves, joints and spinal disks.

## WHAT YOU CAN DO TO PREVENT INJURY

- Develop an Ergonomics program
- Take proper breaks.
- Health and fitness.
- Be aware of your hobbies away from work.

### Ergonomic Breaks

Time taken to give your body including the eyes rest before continuing a job. It relaxes muscles and minimizes red flags.

## OFFICE ERGOMOMICS

By applying ergonomic principles to the office setting, risk factors are minimized, productivity is increased and overall workplace quality is improved. The workstation must be adjusted to promote a neutral position while a person works. When adjusting a workstation, keep in mind that all of the equipment interacts. Making one adjustment may alter another.

## ADJUSTING THE WORKSTATION

1. Adjust the chair
2. Adjust reach requirements
3. Adjust focal requirements

## *CORRECT THE ENVIRONMENT*

1. Check lighting, noise and temperature
2. Check work pace and stress levels
3. Check work process

## *IMPROVE POSTURE AND HABITS*

1. Modify wrist/hand motions
2. Improve neck and back postures
3. Consider personal preference

## *INDUSTRIAL ERGONOMICS*

By applying ergonomics principle in industrial settings, a safer, healthier and more productive work environment can be developed. Employees need to know how to minimize risk factors by choosing the best tools and work techniques for a given task.

## ARRANGE THE WORK AREA

1. Consider the base of support
2. Place equipment and materials where appropriate

## CHOOSE THE APPROPRIATE TOOLS

1. Check the fit
2. Make sure the tools match the task
3. *IMPROVE WORK TEACHNIQUES AND HABITS*
4. Improve postures
5. Check work techniques

## CONTROL STRATEGIES IN ERGONOMICS

The next step is to develop and implement control strategies to increase quality and productivity. Once the risk factors and their causes are identified, control strategies can be implemented based on needs.

### Engineering Controls

1. Appropriate initial design of the work station or work area
2. Improving the design of the existing work area or equipment
3. Providing the work station layout and equipment.

## ADMINISTRATIVE CONTOL

1. Training workers in work methods
2. Varying or rotating work tasks
3. Limited extended work hours
4. Providing mini-breaks (ergonomic breaks)

## Fire Fighting
### Introduction

Fire is a very rapid chemical reaction between oxygen and a combustion material, which results in the release of heat, light, flames and smoke. Or

A combustion process of oxidation characterized by the production of heat, flame and smoke.

Fire outbreak remains one of the dangerous and costliest experiences one can ever encounter. Fire is generally composed of three elements; fuel, oxygen and heat. Despite the best effort of fire preventive/protective measures to control or limit fire damage should therefore form part of our daily operational routine.

## THE FIRE TRIANGLE

Three things are required in proper combination for ignition and combustion to occur:

- Enough OXYGEN to sustain combustion
- Enough HEAT to reach ignition temperature
- Some FUEL or combustion material

**Terms in Fire Fighting**

THE FIRE TRIANGLE

## Heat Transmission

Whenever temperature differences exist between two bodies, heat is transferred. Heat transfer could be by convection, conduction or radiation.

It involves the transfer of heat by the missing and motion of macroscopic portions of a fluid. Boiling water transfers heat through convection.

### CONDUCTION

This is the transfer of heat by the interactions of atoms or molecules of a material through which the heat is being transferred. It is transmitted through metal object or through unprotected steel work; walls, floors, beams steel girders and deck plating.

### RADIATION

It involves the transfer of heat by electromagnetic radiant waves that arises due to the temperature of the body.

## *IGNITION TEMPERFATURE*

Ignition temperature is the lowest temperature at which then substance will ignite spontaneously.

## *OXYGEN*

Oxygen constitutes about 21% of the total volume of air. It is a great supporter of life and combustion. Most fires draw their oxygen from air. Some materials, however, contain enough oxygen in a form that is liberated adequately to support combustion.

## *HEAT*

Heat is a form of energy, and flows from an area of high temperature to an area of low temperature.

Heat source of heat can be from an open flame, electricity, hot surfaces, friction, etc. After ignition temperature has been reached.

Burning will continue as long as fuel remains. The amount of heat produced during a fire is called heat of combustion. All substance at flash point temperature will ignite. This is the lowest temperature at which a substance gives off sufficient vapor to ignite.

## FUEL

Fuel is anything that can burn. It could be solid, liquid or gas. Fuel may exist in these forms, but combustion normally occurs when fuel is in the gaseous or vapor state. Solid and liquids therefore must have applied energy usually heat, to vaporize them before oxygen can react with the fuel in the combustion process.

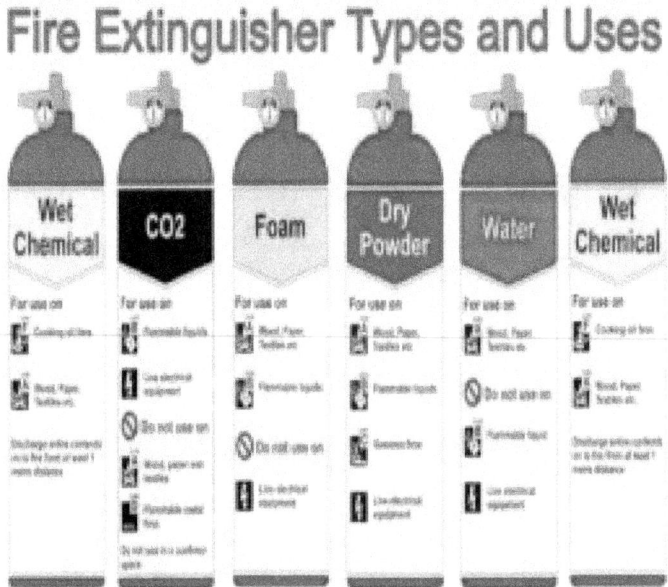

These are different extinguishers for different uses

You do not use the one containing water to extinguish flammable liquids

*WATER*

Large fire extinguishers that stand about 2 feet tall and weigh about 25 pounds when full. APW stands for "Air – Pressurized Water".

Filled with ordinary tap water and pressurized air, they are essentially large squirt guns. Water (H2O) is a commonly used to extinguish most fires. It works because it cools the fire to below its ignition temperature. APW's extinguish fire by taking away the "heat" element if the fire triangle. APW's are designed for Class a Fires only; Wood, paper, cloth.

Using water on a flammable liquid fire could cause the fire to spread. using water on an electrical fire increases the risk of electrocution. If you have no choice but to use an APW on an electrical fire, make sure the electrical equipment is un-plugged or de-energized.

Do not use water extinguishers on electrical equipment which is live or plugged in. It is important to remember, that ELECTRICAL EQUIPMENT must be disconnected from its electrical source before using a water extinguisher on it.

## FOAM

Used for pitting away fire from highly flammable materials. E.g. fire from petrol, Diesel and kerosene.

## CARBONDIOXIDE (CO2)

Carbon dioxide is a non – flammable gas that takes away the oxygen element of the fire triangle. Without oxygen, there is no fire.

$CO_2$ is very cold as it comes out of the extinguisher, so it cools the fuel as well. Carbon dioxide acts to extinguish the fire primarily by dilution and displacement of oxygen.

## Dry- Chemical Powder

Dry-chemical particles inhibit the chemical chain reaction by preventing reactive particles from coming together. It if used on combustible metal fires to smother the fire. Dry chemical extinguishers put out fire by coating the fuel with a thin layer of dust. This separates the fuel from the oxygen in the air. The powder also works to interrupt the chemical reaction of fire.

*These extinguishers are very effective at putting out fire.*

Key facts about fire extinguisher

- *P- PULL THE PIN*
- *AIM THE NOZZLE*
- *S- SQUEEZE THE TRIGGER*
- *S- SWEEP THE*

# CHAPTER 3 HSE LEVEL 3 (SUPERVISORY HSE)

## Environmental Management

The environment is man's life support system. It consists of air, water, land, animals, plants, and human beings. The interrelationship between them is critical to the existence of all. This is the essence of the need to pay attention to protecting the environment as we carry out our activities. The environment is very important for our well-being and existence. the activities of man destroy the natural quality of the environment, directly or indirectly, through impact, pollution, or over exploitation.

*How is the environment affected by our operations?*

- *Operation of equipment*
- *Maintenance of equipment*
- *Waste handling*
- *Environmental awareness*
- *Construction Activities*

1. Your responsibilities are minimizing impacts on the environment
2. Read the HSE policies in your organization
3. Proper waste handling
4. Report an environmental hazard in operations to appropriate authority.
5. Cherish the natural environment
6. Stop environmental unsound activities.
7. Environmental Protection
8. Environmental protection is the protection and conservation of existing resources. It is the act of ensuring that air quality;

water/groundwater quality, soil quality, noise, heat, light, vibration, vegetation, socio-economic/community impact, wildlife, fisheries, etc. are conserved
9. Environmental issues in exploration and production activities/operations
10. Exploration and production operations cover over 50,000 km of the Niger Delta and its immediate waters. These operations include the following.
11. Exploration and appraisal drilling
12. Development and production
13. Other installations / facilities
14. Over 1000 producing wells
15. Over 10,000km of pipelines and flow lines
16. Over 100 flow stations and gas plants

Environmental management involves the protection and conservation of natural resources, from the air we breathe, to the water we drink to the ecosystems that support life to the renewable and non-renewable energy and materials that are exploited for modern living and the disposal of waste.

## Principles Of Environmental Health And Safety

All the tragedies in the workplace can be avoided by sound prevention, reporting, and inspection practices. The global strategy to improve occupational safety and health adopted in 2003 by the ILO included the introduction of a preventive safety and health culture, promotion and development of relevant instruments, and technical assistance.

The ILO has adopted more than 40 standards, specifically dealing with occupational safety and health, as well as over 40 codes of practice. Nearly half of ILO instruments deal directly or indirectly with occupational safety and health issues (ILO 1981).

The principles of environmental health and safety target the protection and safety of working men and women. In any country, these can be achieved through the setting and enforcing standards, the provision of education, outreaches and assistance, issuing of permits, licenses, certificates, registrations

and approvals.

*The Environment Can Be Subcategorized into:*

- Air emissions and ambient air quality
- Energy conservation
- Wastewater and ambient water quality
- Water conservation
- Hazardous materials management
    - Waste management
    - Noise
    - Contaminated land
- The occupational health can be subcategorized into:
    1. General facility design and operation

2. Communication and training
3. Physical hazards
4. Chemical hazards
5. Biological hazards
6. Radiological hazards
7. Personal protective equipment (PPE)
8. Special hazard environments
9. Monitoring

- Community health and safety can be subcategorized into:
1. Water quality and availability
2. Structural safety of project infrastructure
3. Life and fire safety (L&FS)
4. Traffic safety
5. Transport of hazardous materials
6. Disease prevention
7. Emergency preparedness and response

The Environmental Health and Safety Principles Can Be Discussed Under the Following Principles According to Knoll, 2014.

*1. Protection of the Biosphere*

To make continued progress toward reducing or eliminating the release of any hazardous substance in an effort to safeguard all habitats affected by our operations" This principle can be achieved by:

The reduction of the use and/or emissions of hazardous air pollutants and volatile organic compounds from manufacturing operations through the introduction of clean technologies.

The provision of water treatment facilities that meet or exceed discharge criteria.

The monitoring of storm water, conserve water use and develop processes to efficiently use water and minimize water pollution.

## 2. Sustainable Use of Natural Resources

To make the best use of renewable resources, such as water, soil and forests, and conserve nonrenewable resources. This principle can be achieved by:

Sustainable use of renewable natural resources through efficient use and careful planning.

Seeking opportunities to use wood from sustainable forests in products.

An attempt to recycle or make beneficial use of wood scrap generated in manufacturing operations.

Recycling steel, aluminum and other metal components.

Being environmentally responsible in the purchase of materials.

*3. Waste Reduction and Disposal*

To reduce, recycle, and where possible, eliminate waste and disposing of all waste using safe and responsible methods with the intention of eliminating the landfilling of waste" This principle can be achieved by an attempt to seek opportunities to reduce waste and introduce recycling process in an organization's operations.

Disposing of wastes only in well-operated and permitted facilities.

*4. Conservation*

To conserve energy by improving the efficiency of internal operations.

***This principle can be achieved by:***

- Making every effort to use environmentally safe and sustainable energy sources.

- Conserving energy and improve energy efficiency.

## 5. Risk Reduction

"To strive to minimize the environmental health and safety risks to employees

and the communities in which the industries operate through safe technologies, sound transportation practices, safe facilities and operating procedures, and preparing for emergencies" This principle can be achieved through:

The designing of processes to prevent injury to the health and welfare of workers, the community and the environment.

The development and implementation of health and safety policies and programs to help prevent injury and illnesses to workers.

The development and implementation of health and wellness awareness and illness prevention programs.

Designing and developing training programs to provide workers with the necessary skills and knowledge to fulfill the objectives of the

Environmental, Health and Safety Plan.

## 6. Safe Products and Services

To reduce and, where possible, eliminate the use, manufacture or sale of products and services that cause environmental damage or health or safety hazards. This principle can be achieved through:

The provision of independent testing to assure the safety of products.

### 7. Environmental Restoration

To comply responsibly with the law to address conditions caused by the industrial process that could endanger health, safety or the environment.

### 8. Public Information

To comply with the law to inform on a timely manner those who may be affected by conditions caused by operations that might endanger health, safety or the environment and will encourage employees to report dangerous incidents or conditions to management.

The Following Principles Must Be Applied Every Day and at All Sites in Order to Succeed in Our Commitment

- Remove or alleviate hazardous situations in order to minimize risks to personnel

- Set HSE performance objectives, measure results, assess and continually improve processes, through the use of an effective management system

- Monitor the health of our personnel on a regular basis to ensure that workers have the physical ability to perform their duties

- Guarantee and monitor that our activities have minimal impact on the environment.

- Minimize our impact on the environment through pollution prevention, reduction of natural resource consumption and emissions, and the reduction and recycling of waste

- Apply our technical skills to all Health, Safety and Environment aspects in the design and engineering of our services and products

- Promote awareness of issues relating to Health, Safety and Environment towards our employees and contractors, improving their individual skills through appropriate training programs and information.

- Develop, maintain and test procedures and resources for dealing with emergencies in order to reduce the impact of incidents to a minimum

- Report and analyses all accidents, incidents and anomalies.

- Monitor our performance and check our management system in order to improve rules and procedures in conducting our activities.

- Maintain a consistent and transparent dialogue with authorities, local realities and other stakeholders, openly communicating our HSE policies, standards, programs and performance

## Health Safety And Environment Supervision

### *What Is Supervision*

Supervision is the act of guiding and directing someone or a group of persons on what to do and what not to. In general, supervision is the act of seeing and supervising the overall project of an organization.

Health, safety, and environmental (HSE) supervision is a critical component of industrial and organizational management practices aimed at ensuring the well-being of workers, protecting the environment, and complying with regulatory standards. This supervision involves comprehensive monitoring, evaluation, and implementation of protocols to minimize risks and promote sustainable practices across various industries.

## *Importance of HSE Supervision*

HSE supervision plays a pivotal role in safeguarding the health and safety of employees within workplaces. It involves conducting risk assessments, implementing safety procedures, and providing training to mitigate occupational hazards such as chemical exposures, ergonomic issues, and physical injuries. By fostering a culture of safety and compliance, HSE supervision helps reduce workplace accidents and illnesses, thereby improving workforce productivity and morale.

## *Regulatory Compliance*

HSE supervision ensures that organizations comply with local and international regulations governing occupational health and environmental protection. Regulatory bodies, such as the Occupational Safety and Health Administration (OSHA) and the Environmental Protection Agency (EPA), set standards for workplace safety, chemical handling, and emissions control (OSHA, nod). Compliance with these regulations is essential to avoid legal liabilities, fines, and reputational damage.

## *Continuous Improvement and Training*

Continuous improvement is integral to effective HSE supervision. Regular audits, incident investigations, and feedback mechanisms enable organizations to identify weaknesses and implement corrective actions. Employee training and awareness programs also play a crucial role in promoting a safety-conscious culture and ensuring compliance with HSE protocol.

In conclusion, HSE supervision is indispensable for promoting health, safety, and environmental stewardship in industrial and organizational settings.

By integrating proactive monitoring, regulatory compliance, and continuous improvement initiatives, HSE supervision helps mitigate risks, protect resources, and enhance overall organizational performance. Emphasizing the importance of HSE supervision fosters a sustainable approach to business operations that prioritizes the well-being of workers, the environment, and society at large.

### Here Are 40 Duties Of A Safety Officer

1. The Safety Officer is responsible for monitoring and assessing hazardous and unsafe situations.

2. Developing measures to assure personnel safety.

3. Correct unsafe acts or conditions through the regular line of authority.

4. May exercise emergency authority to prevent or stop unsafe acts when immediate action is required.

5. The Safety Officer maintains awareness of active and developing situations.

6. Ensures there are safety messages in each Incident Action Plan.

7. Participate in planning meetings to identify any health and safety concerns inherent in the operations daily work-plan.

8. Review the Incident Action Plan for safety implications.

9. Investigate accidents that have occurred within incident areas.

10. Ensure preparation and implementation of Site Safety and Health Plan (SSHP).

11. Inspects the site to ensure it is a hazard-free environment.

12. Conducts toolbox meetings.

13. A HSE Officer is part of the project safety council and leads all efforts to enhance safety.

14. The safety officer reviews and approves all subcontractor's safety plans.

15. Verifies that injury logs and reports are completed and submitted to related government agencies.

16. Verifies that all tools and equipment are adequate and safe for use.

17. Promotes safe practices at the job site.

18. Enforces safety guidelines.

19. Trains and carries out drills and exercises on how to manage emergency situations.

20. Conducts investigations of all accidents and near-misses.

21. Reports to concerned authorities as requested or mandated by regulations.

22. Conducts job hazard analysis.

23. Establishes safety standards and policies as needed.

24. Watches out for the safety of all workers and works to protect them from entering hazardous situations.

25. Responds to employees' safety concerns.

26. Coordinates registration and removal of hazardous waste.

27. Serves as the link between state and local agencies and contractors.

28. Receives reports from and responds to orders issued by Department of Labor.

29. Arranges for OSHA mandated testing and/or evaluations of the workplace by external agencies/consultants.

30. Support the development of OHS policies and programs.

31. Advise and instruct on various safety-related topics (noise levels, use of machinery etc.).

32. Conduct risk assessment and enforce preventative measures.

33. Review existing policies and measures and update according to legislation.

34. Initiate and organize OHS training of employees and executives.

35. Inspect premises and the work of personnel to identify issues or non-conformity (e.g. not using protective equipment)

36. Oversee installations, maintenance, disposal of substances etc.

37. Stop any unsafe acts or processes that seem dangerous or unhealthy

38. Record and investigate incidents to determine causes and handle worker's compensation claims

39. Prepare reports on occurrences and provide statistical information to upper management.

40. Carry out PTW Monitoring and review

## Harzard And Effect Management Process In Safety

The Hazard and Effect Management Process (HEMP) is a systematic approach used in safety management to identify, assess, and mitigate risks associated with hazardous activities or environments. It is commonly used in industries such as aviation, healthcare, manufacturing, and oil and gas, among others.

### *The Process Typically Involves Several Key Steps:*

***Hazard Identification:*** The first step is to identify all potential hazards within a system, process, or environment. This can be done through various methods such as brainstorming sessions, historical data analysis, inspections, and using checklists.

***Risk Assessment:*** Once hazards are identified, they are assessed to determine the likelihood and severity of their potential effects. This involves analyzing the probability of the hazard occurring and the potential consequences if it does. Risk assessment techniques such as risk matrices, fault tree analysis, and event tree analysis may be used.

*Risk Control:* After assessing the risks, control measures are implemented to reduce or eliminate them to an acceptable level. This may involve engineering controls, administrative controls, or procedural controls. The goal is to prevent accidents, injuries, or damage to property or the environment.

*Monitoring and Review:* The effectiveness of the control measures is monitored and periodically reviewed to ensure they remain adequate and effective. This may involve conducting regular inspections, audits, and safety assessments.

*Documentation and Communication:* All findings, assessments, control measures, and reviews are documented for record-keeping and communication purposes. This ensures that relevant stakeholders are informed and aware of the hazards and associated risks.

*Continuous Improvement:* The HEMP process is iterative, and continuous improvement is essential. Lessons learned from incidents, near misses, and feedback from stakeholders are used to refine and improve the safety management system over time.

HEMP is often implemented alongside other safety management systems such as Safety Management Systems (SMS) or Risk Management Systems (RMS) to ensure comprehensive hazard and risk management. It helps organizations proactively identify and address safety concerns, leading to safer operations and reduced incidents and accidents.

## Challenges Affecting The Effective Implementation Of Safety

***Limited Resources:*** Safety centers often lack adequate funding, personnel, and equipment necessary to effectively carry out their mandate. This scarcity of resources can hinder the provision of essential safety services and infrastructure.

***Infrastructure Deficiency:*** Many safety centers in Nigeria operate in facilities that are poorly maintained or lack basic amenities such as electricity and water supply. This can impede the delivery of timely and efficient safety services to the public.

***Lack of Awareness:*** There is often a low level of awareness among the populace regarding the existence and functions of safety centers. This lack of awareness can lead to underutilization of safety services and a failure to report safety concerns or emergencies promptly.

***Inadequate Training and Capacity Building:*** Staff members at safety centers may lack the necessary training and skills to handle safety-related issues effectively. Continuous training and capacity-building initiatives are essential to ensure that personnel are equipped to respond to safety challenges competently.

***Bureaucratic Bottlenecks:*** Complex bureaucratic procedures and administrative hurdles can slow down decision-making processes within safety centers, hampering their ability to respond swiftly to safety incidents and emergencies.

***Political Interference:*** Safety centers may face political interference, which can compromise their autonomy and ability to function effectively. Political influence may affect resource allocation, staffing decisions, and operational priorities, undermining the overall effectiveness of safety initiatives.

***Security Challenges:*** In some parts of Nigeria, safety centers may face security threats such as vandalism, theft, or attacks by criminal elements. Ensuring the security of personnel and facilities is crucial for the uninterrupted provision of safety services.

***Corruption:*** Corruption within safety centers can undermine efforts to promote safety and security. Mismanagement of funds, bribery, and other corrupt practices can divert resources away from essential safety initiatives, weakening the overall effectiveness of safety centers.

Addressing these challenges requires concerted efforts from government authorities, civil society organizations, and other stakeholders to provide adequate resources, improve infrastructure, raise awareness, enhance training programs, streamline administrative processes, safeguard against political interference and corruption, and ensure the security of safety centers and their personnel.

## Waste Management

### WHAT IS WASTE?

A waste is a material that no longer serves its original purpose and will be discarded. Waste is anything that is no longer of use to the disposer. Any unavoidable material resulting from an activity, which has no immediate economic demand and which must be disposed of.

### FORMS OF WASTE

Wastes, no matter where it is generated, appears in three forms. They are:

- Gaseous wastes
- Liquid wastes
- Solid wastes

## GASEOUS WASTES

Wastes such as particulate dust, waste gases from stack, cement factories, gas flaring, stone crushing excavation activities, lime dust, asbestos dust, acid fumes and cigarette fumes.

## LIQUID WASTES

Examples include waste dissolved in water emanating from industrial processes, domestic liquid, acid waste, and waste oil from workshop.

## SOLID WASTE

Waste such as broken bricks, refuse, sludge and slag, broken glass and bottles, can, plastics, paper, battery casings, plantain skin, pure water, etc.

## CLASSES OF WASTE

Before you can properly manage waste, you must know something about it. Waste management is important in order to minimize the potential to cause harm to human and the environment. Waste classification is the process by which a waste is identified as a domestic or industrial waste.

- Domestic or municipal waste
- Industrial waste

## DOMESTIC WASTE

Examples of domestic wastes are garden waste, office waste kitchen waste, paper, food, wood, laundry. Waste water from toilet and bathrooms. They can also be called general trash. General trash includes paper cups. food scraps. General trash does not include hazardous waste, waste contaminated with process material, incandescent light bulbs, fluorescent light bulbs, Ni-Cad batteries, aerosol cans, scrap metal, liquids and tires.

## *INDUSTRIAL WASTES*

Wastes such as oil waste, effluents or gaseous emissions (smoke, fumes, particulate dust), scrap metals, cardboard packaging, plastics, organic acid, chemical waste.

## *HAZARDOUS WASTE*

A hazardous waste is a waste that has the potential to harm human health and the environment if not handled properly, e.g. Corrosive ignitable/explosive. Hazardous waste is to be treated before disposing them.

## NON-HAZARDOUS WASTE

A non-hazardous waste is a waste that may not pose a serious threat to human health or the environment but may still be regulated. They are environmental friendly and biodegradable.

## WASTE PROFILE

A waste profile is a detailed written description of the waste stream listing the stream's characteristics (i.e. lists components, range, physical state, color, odor, etc.).

## E&P EXEMPT WASTE

Waste directly generated from production operations. E.g. produced water, pit sludge, produced sand, spent glycol, amine and gas plant dehydration & sweetening wastes.

## NON-CLASSIFIED WASTE LABEL

A non-classified waste is any waste that has not yet been classified as either hazardous or non-hazardous. Wastes are classified for the following reasons:

- Identification of wastes with hazardous potential
- Identification of hazardous waste
- Waste segregation
- Regulatory requirements.

## WASTE MANAGEMENT PRINCIPLES

These are proven principles used in managing waste.

***They are waste:***

- Inventory
- Characterization
- segregation

- Minimization
- Treatment

## Inventory

This is the Cataloguing of all waste types, quantities and sources. It gives a quick insight on the magnitude of the waste problem at hand.

## CHARACTERIZATION

It has to do with checking waste properties such as physiochemical and toxicological properties.

## SEGREGATION

Optimal recovery of waste streams through selective separation of waste.

## MINIMIZATION

The reduction, reusing, recycling or recovering of waste.

## TREATMENT

Physical/biological treatment of wasted to reduce it toxicity.

## Waste Disposal Options

Waste handling important because it keeps you and others around safe. Proper handling of waste ensures that human health and the environment are protected and it is the law. Resources Conservation and Recovery Act (RCRA) is the federal regulation covering hazardous waste management and disposal and is enforced by the Environmental Protection Agency (EPA). Waste can be disposed through the following ways:

- Re-injection
- Solidification, Encapsulation (compress into solid)
- Bio-treatment
- Surface discharge/open dump
- Ocean dump
- Thermal treatment or incineration
- Landfill

## LANDFILL

Land fill means digging the ground, dumping the waste and covering it. Biodegradable wastes are buried in landfill. Hazardous wastes such as nuclear reactors are buried at a distance of 500m below the earth surface. The three types of landfill are:

## INERT LANDFILL

Non decomposable, non-water soluble wastes e.g. nuclear wastes which is buried at a distance of 500m below the surface of the earth and monitored.

## SANITARY LANDFILL

For non-hazardous wastes, Biodegradable material, municipal waste/industrial waste.

## CHEMICAL LANDFILL

Used for hazardous wastes from industries pre-treatment before disposal.

## POOR WASTE MANAGEMENT HAZARDS

Environment with poor waste manage practices are always unsightly. The following hazards are likely to be present in such place:

- Unaesthetic dump site
- Foul odors-loss of community pride
- Leach ate from dumpsites can poison our surface and ground water.
- Stagnant pools provide breeding grounds for mosquitoes, flies and diseases vector.
- Provides abundant food for rodents which transmit harmful bacteria and virus leading to epidemic – Ebola, Lassa, Plague
- Fire outbreak.

## WHY LABEL WASTE CONTAINERS

Certain information is required on containers of waste for efficient segregation. Most industries accomplish these requirements with labels. Waste containers are labeled to:

- Keep you and others safe
- Identify what waste is in the container
- Identify the hazards present
- Avoid regulatory penalties and government action

## WHERE DO I PUT MY WASTE?

- Waste must be placed into a container that:
- Can hold the waste safely
- Is approved by the receiving facility
- Is clean and in good condition

## Different Types Of Waste Containers

### BULK CONTAINERS

These are container that can hold more than 119 gallons of waste. An example includes roll-off bins, tank trucks, rail cars and tote bins.

### NON-BULK CONTAINERS

Non-bulk containers hold 119 gallons of waste or less. Examples are drums, boxes, pails, totes.

*None bulk waste container.*

**Industrial Security Management**
*Introduction*

Security is every body's business. Security is protection of organizational assets against identified risk such as spying, attacks, theft, sabotage and vandalism.

it is also the science or art of ensuring the safety of lives and property entrusted into one's care at all times by means of prevention, protection and preservation.

## SECURITY RELATED DEFINITIONS

The following is a compilation of working definitions for various security terms.

## *THREAT ASSESSMENT*

Identification and analysis of specific incidents, criminal activity, political, social, economic and environmental factors that precipitate issues and events that could potentially pose threats to particular company personnel, assets, operations, facilities, reputation, and business information.

## *THREAT STATEMENT*

It is a document that is developed from a review of applicable potential security threats, and focuses on a specific company activity and or location.

## *SECURITY RISK MITIGATION*

The process of identifying, evaluating, and stewarding implementation of appropriate countermeasures necessary to control, eliminate or minimize security risks, and review of countermeasures necessary to control, eliminate or minimize security risks, and review of countermeasure effectiveness.

## *SECURITY SURVEY*

A structured and detailed security assessment based on the application of risk management principles. The ultimate goal is to improve the security program by identifying threat-driven exposures and recommending cost-effective security countermeasure.

## *SECURITY AWAERNESS TRAINING*

Providing education and awareness programs to inform personnel and family members, as appropriate, of potential exposures to security risks, accountability for the protection of company assets, and measures individuals can take to mitigate these risks.

## SECURITY HAZARD

Security hazards are physical or procedural deficiencies that create a potential exposure to a security threat. Examples could include a gate or door lock that doesn't positively lock always, a gap that's developed under a perimeter fence, brush overgrowth in perimeter area that prevent full surveillance by CCTV.

## SECURITY NEAR MISS

Similar to safety near miss, these are situation which under slightly different circumstances could have exposed operations or personnel to a security threat.

## BASELINE SECURITY PROTECTION

Recommended minimum considerations for security countermeasures. Application of specific security countermeasures is based on security threat/risk and cost-benefit analysis.

Once an asset is identified as requiring protection, minimum considerations must be given to basic security principles of detection, delay and response, with additional measures added based on security threat and risk. See ISMS Appendix R.

## *ALARM SYSTEMS*

Sensor systems that provide a high probability of detection into controlled areas, and duress alarms.

## *ACCESS CONTROL*

Application of ingress, egress, circulation controls, as well as procedural and physical barriers to restrict movement to authorized individuals within EM facilities and to provide graduated concentric zones of security.

## *CCTV*

Close circuit television or other remote video/computer/digital image surveillance methods used to assess, respond, and record events in areas of security interest.

## *EXPLOSIVES/WEAPONS DETECTON*

Explosive and weapons identification and detection systems and related munitions countermeasures and controls, such as z-ray screening or metal detectors.

## *PHYSICAL SECURUTY MEASURES*

This is the placement of checks in processes or security operatives at strategic positions to achieve a given objective.

## TYPES OF PHYSICAL SECURITY MEASURES

- Access control
- Investigation
- Security patrol
- Intelligence/Surveillance

## ACCESS CONTROL

Strategic positioning of competent personnel with transparent character at entry points to a site or work location in order to control entries and exits. This is achieved by use of a some of the following;

- Enforcement of wearing valid ID cards
- Movement monitoring with CCTV at access gates
- Visitor screening and control/ reception
- Use of physical barriers

## INVESTIGATION

Technically used to investigate/know the cause nature and dimension of crime after an incident in order to control future occurrence.

## SECURITY PATROL

This is a measure used to deter crime by tracking all unauthorized entrances.

### Patrol Types

There are four types of patrol in security, namely:

- Fixed route
- semi-discretionary system
- Discretionary system
- Aberdeen

## INTELLIGENCE \ SURVEILLANCE

Discreet monitoring of activities within an installation, asset, or site to prevent crime and banalization.

## SECURITY INCIDENT RECORDING AND REPORTING

### Security Incident

Security incident is an event whereby the willful actions of one or more person injure, threaten the safety of, or endanger other. Each incident may affect one or more people and /or properties.

- It involved a personnel or Company property
- It is a New case
- It is Work- Related
- It meets one or more of the security Incident definitions

## Security Reporting

Report threatening and / inappropriate behavior or suspicious person or activities such as:

- Unknown person photographing facilities
- Unauthorized person (s) trying to gain access to facilities
- Suspicious package drop-offs or attempted drop- offs
- Repeated or out of ordinary phone calls
- Attempts are made to obtain information concerning the facility, facility personnel or facility operations

## TYPES OF SECURITY INCIDENTS

The following are security incidents and are corporate reportable:

- Arson
- Actual Assault
- Break – in
- Burglary
- Homicide
- Kidnapping
- Public Disturbance / Activism
- Robbery
- Theft
- Sabotage
- vandalism

### Arson

Intentional and malicious burning to destroy company property

*Assault*

Any attempted or actual infliction of verbal or physical injury on a person.

*Break – In:* Entering company property by force.

*Burglar:* Theft of company property involving forceful, unauthorized entry into a building or structure.

## HOMEICIDE

The killing of one person by another.

## KIDNAPPING

The forcible detention, abduction or confinement of company personnel. Kidnapping includes:

*Express Kidnapping:*

A type of short duration kidnapping that generally involves forcing the captive to withdraw money from a particular banking account.

*For Ransom:* A kidnapping that involves a request for money or a demand that must be fulfilled in return for the release of the captive.

*Robbery:* The taking of company property with the use of force, threat of force, or violence against the victim. Examples include carjacking, marine vessel piracy and truck jacking.

*Theft:* The taking of company property without permission, e.g. embezzlement and extortion.

*Threat:* An expression of intention to inflict injury or damage. e.g. a threat to detonate an explosive (Bomb threat)

*Sabotage:* Willful and malicious actions in order to destroy company property, interfere with, or adversely affect normal operations.

*Trespassing:* Evidence of unauthorized intrusion or invasion of company property, irrespective of resulting damage.

*Vandalism:* Malicious acts intended to damage or destroy company property.

### Indentity Theft

A crime in which someone wrongfully obtains and uses another person's personal data in some way that involves fraud or deception, typically for economic gain. e.g. being defrauded of funds from your financial accounts. The cost of trying to repair the damage that has been done is always high. Identify theft can be in the following form:

- Personal information
- Social security or other government ID number
- Credit card numbers
- Telephone calling card number
- Passwords or PINs
- Physical Environment
- Mailbox
- Phone
- Trash- "dumpster diving"
- ATM – "shoulder surfing"

- Online Environment
- Ordering with credit cards
- Emails from strangers
- Not logging out of sessions
- Accessing confidential information on public computers
- Phishing scams

*What can you do?*

- Avoid giving out personal data
- Do not open emails from unknown sources
- Shred receipts and other private documents
- Memorize pins
- Check your bank accounts and credit records often
- Maintain careful records of your financial accounts.

## FACILITY SECURITY

- Access Control and Technical Security
- Park in designated facility parking areas
- utilize the facility's sign-in /sign-out procedure (E.g. card readers, log sheets, photo identification)
- Do not allow others to "piggyback" or "tailgate" into the facility
- Escort Visitors
- Professionally and politely challenge unescorted visitors
- Where issued, wear company ID badges
- Report lost or stolen ID badges immediately
- Know the facility's emergency alert and security response procedures
- Refer to site specific emergency plans
- Field areas utilize security checklists to implement countermeasures during elevated threat levels (i.e. blue, yellow, orange, red)

- Bomb threat checklists must be distributed and placed near facility phones

## MANAGEMENT AND PROTECTION OF INFORMATION

This is the restriction of access to sensitive documents with the use of control paints such as locked storage, locked rooms, and card reader controlled room/area with limited access.

## OFFICE INFORMATION SECURITY

### Mail handling procedures

This process whereby all incoming mail is screened by mail handling personnel. Elevated threat levels may require the use of mail screening devices such as x-ray.

Restrict couriers or delivery personnel from making personal deliveries to the facility – except for those who are pre-screened and approved.

1. Information security practice
2. Activate passwords protected screensaver when workstation is unattended
3. Retrieve sensitive documents from fax machines, printers and mailboxes promptly
4. Log off computer at the end of the workday
5. Protect passwords and personal identification numbers (pin)
6. Avoid cordless or cell-phones when discussing sensitive matters
7. Know the sensitivity of the material you are carrying
8. Increase security awareness in high threat environments and during sensitive times
9. avoid being overheard if discussing sensitive information in public places – be aware of

your surroundings, move to a more secure area
10. Inform a supervisor, a security advisor and building security if someone is attempting to obtain sensitive information.

## WORKPLACE SECURITY PRECAUTIONS

1. Always display your identification Badge.
2. Offer to assist visitors or those without proper company identification to the security desk
3. Do not assist unauthorized entry. Report cases of tailgating to security.
4. Reduce opportunity for theft at the office by placing purses and other valuables of a locked cabinet. Keep your wallet with you always or secured in a locked area. Do not leave your wallet, checkbooks on the desk or in an unsecured office.
5. Know what to do in the event of a security emergency.

6. When leaving the office at the end of the workday, please secure your work area. Lock your desk drawers and cabinets. Make sure that all classified materials are secured and all classified envelopes are removed from mail bins. It is important to protect all sensitive materials, company and personal property.
7. If you plan working late, please notify your supervisor and security group in writing. Be sure that your car has been positioned for easy entry in a secure location.
8. In your vehicle, keep all valuables out of sight. Lack cell phones, portable radios and CD players, palmtop computer, and other valuables in the glove compartment or trunk.
9. Immediately report suspicious individuals, suspicious vehicles, unattended packages, briefcases, equipment or other unusual items to security.

## SECURITY AT HOME

1. Properly lock all doors and windows every time you leave the house, even if only for a few minutes.
2. Maintain visible and well-lit entry ways
3. Communicate with neighbors and family members

## LAPTOP COMPUTER SECURITY TIPS

1. Always use laptop tether cables and attach the tether cable to a secure object
2. Lock your work station (laptop), insure the tether is attached and lock the office door when you are out of your office.
3. When traveling, keep laptop computers with you at all times or lock them either in the trunk of your vehicle or in a safe place such as hotel safe.
4. Place laptops inside other carry-on luggage

5. Do not put laptop computers into luggage that is to be checked in
6. Power – on / Hard Drive passwords must be installed and used on company laptop computers.
7. Immediately report the loss or theft of a company laptop computer to your supervisor, the security advisor and or the site security contact.

## Personal Security Measures

Security is everyone's responsibility. Your personal security is very critical to you first and foremost. As an employee, you are expected to have your own security plan. This will ensure that the effort of the security experts (our company's security department or that of the facility owner) is complemented.

1. The following are useful tips that could assist you in formulating your own security plans:
2. Understand the possible threats in the work environment. Ask questions where in doubt.
3. Develop a sense of security awareness. Do not take security for granted
4. avoid routine patterns of activity as much as possible
5. Protect personal information
6. Adopt realistic security measures
7. Have contingency security plan
8. Avoid self-targeting

9. Avoid unnecessary exposures to risks
10. Understand and obey security rules and regulations of your company.

## General Security Guidelines

1. Take time to plan activities. Try to know the exact route before traveling
2. Dress ad behave appropriately, gibing consideration to local customs
3. Learn a few words or phrases in the local language to deter and offender or call for help, such as "police" or "fire"
4. At a new assignment, find out about local customs and behavior and potential threats or areas to avoid.
5. Know the local security arrangements, such as the nearest police station, Emergency contact procedures, and potential safe areas
6. Maintain a calm, mature approach to all situations
7. Be non-provocative when confronted with hostility or potentially hostile situations
8. Be alert to the possibility of confrontation with individuals or groups. Be aware of times

when crowds can be expected, such as after religious services or sporting events

9. All international staff, family members, and visitors should register with their embassy or consulate. They should know the telephone numbers, contact personnel, location and emergency procedures for their embassy.

### Travel Security

- Use hard case, lockable luggage and label it so the name and address are not easily seen.
- When traveling, leave a planned itinerary with a responsible person.
- Carry a list of emergency names, addresses, phone numbers, and the names of reputable hotels along the route.
- When appropriate, photocopy passport and other documents and carry only the copy, keeping a second copy at home or office.

When carrying the original, consider disguising it with a plain slip-on cover.
- Organizations should provide photo identification cards for all staff and emergency contact cards for visitors. They can be laminated, two-sided cards with English or another UN standard language on one side and the local official language on the reverse.
- Carry a phone card or local coins to make emergency phone calls if required.
- In public areas or on local transport, sit near other people and hold all belongings.
- Use caution when taking taxis in areas where cab drivers are known to be involved in criminal activity. When available, take licensed taxis and always settle on the fare before beginning the trip. Have the destination address written out in the local language to show the driver if necessary.

## *WALKING*

1. In most setting it is possible to walk safely to and from work or on errands. Walking can help
2. Increase exposure to the community and build acceptance, dispelling the image of the privilege aid worker taking a vehicle everywhere. When the situation permits walking, the following precautions should be put in place:
3. Seek reliable advice on areas considered safe for walking. Consult a local street map before leaving and bring it along.
4. Be aware of surroundings. Avoid group of people loitering on the streets.
5. If possible, walk with companions.
6. Avoid walking too close to bushes, dark doorways, and other places of concealment
7. Use well-traveled and lighted routes.

8. Maintain a low profile and avoid disputes or commotion in the streets.
9. Never hitchhike or accept a ride from strangers.
10. If a driver pulls alongside to ask for direction, do not approach the vehicles. A common criminal technique is to ask a potential victim to come closer to look at a map.
11. Carry all belongings in a secure manner to prevent snatch – and –run theft.
12. If someone suspicious is noted, cross the street or change directions away from them. If necessary, cross back and forth several times. If the person is following or becomes a threat, use whatever means necessary to attract attention of others.

### Public Transportation

1. Avoid traveling alone.
2. Have the proper token or change ready when approaching the ticket booth or machine.
3. During off-pack hours, wait for the train or bus in a well-lit, designated area
4. Be mindful of pickpockets and thieves when waiting for transportation
5. If bus travel at night is unavoidable, sit near the driver. Avoid riding on deserted trains or buses.
6. If train travel at night is unavoidable, select a middle car that is not deserted and try to sit by a window. This provides a quick exit in the event of an accident. Leave any public transport that feels uncomfortable or threatening. After getting off any public transport, check to be sure no one is following.

### Criminal Activity

In recent years, criminal activity has become a significant threat to the safety of aid workers. Criminal activity can take many forms, including armed assault, hijackings, or robbery. Be aware of the extent and activities of organized crime and take necessary precautions (for detailed information in dealing with specific incidents, see (chapter five – Safety and Security Incident). General precautions against criminal activity include:

1. Avoid tourist area that are often favorite places for criminal activity
2. Do not display jewelry, cash, keys, or other valuables in public.
3. Pickpockets often work in pairs using distraction as their basic ploy. BE aware of jostling in crowded areas.
4. When carrying a backpack or purse, keep it close to the body. Do not carry valuables in these bags; instead, leave them in a secure place.

5. It is better to carry only a small amount of money and a cheap watch to hand over if threatened. Divide money and credit cards between two or three pockets or bags.

### Hotel Security

1. Be sure the hotel is approved by the organization. If possible, contact the appropriate embassy for security and evacuation information for that location.
2. Take note if people are loitering in front of the hotel or in the lobby. Avoid hotels frequented by criminals.
3. Ask for a room between the second and seventh floors, avoiding the top floor. This minimizes unwanted access from outside the building yet is within reach of most fire-fighting equipment.
4. Be alert to the possibility of being followed to the room.

5. Advise colleagues of hotel location and room number.
6. Note the evacuation route in case of fire or emergency. Keep a flashlight by the bed to aid emergency evacuation.
7. Always secure doors when inside the room, using locks and security chains.
8. Examine the room, including cupboards, bathrooms, beds, and —window areas for anything that appears suspicious.
9. If the room has a telephone, check to be sure it is working properly.
10. Keep room curtains closed during hours of darkness.
11. Do not open the door to visitors (including hotel staff) unless positively identified. Use the door peephole or call the front desk for verification.
12. When not in the room, consider leaving the light and TV or radio on.

13. If available, use the hotel's safe deposit boxes for the storage of cash, traveler's checks, and any other valuables. Do not leave valuables or sensitive documents in the room.

### Security At Check Points

Checkpoints are manned by personnel with varying degrees of experience, education, or training. Regard all checkpoints with caution, especially in the evening. All staff should receive specific training on identifying and navigating the variety of checkpoints encountered in a given area.

1. Avoid check points whenever possible. Increase attentiveness when approaching checkpoint or possible threat areas.
2. Consider later departure times to ensure others have traveled the route. When approaching a checkpoint or threat area, if possible allow others to pass through the area and observe from a safe distance.

3. Approach slowly with window slightly opened.
4. At night, switch to low beams and put on the interior light.
5. Be ready to stop quickly, but stop only if requested.
6. Keep hands visible at all times. Do not make sudden movements.
7. Show ID if requested, but do not surrender it unless it is insisted.
8. Leave the vehicle only it requested. If the checkpoint is not judged to be an attempted carjacking, turn the vehicle off and take keys. Remain close to the vehicle if possible.
9. Do not make sudden attempts to hide or move items within the vehicle. High theft items, such as radios, cameras, and computers, should always be stored in nondescript containers or kept out of sight.

10. Comply with requests to search the vehicle. Accompany the searcher to ensure nothing is planted or stolen.
11. Use judgment about protesting if items are removed. Do not aggressively resist if something is taken. Request documentation if possible.
12. Do not offer goods in exchange for passage. This can make it more difficult for later travelers.

## *Ultimate 'S Health and Safety Policy*

We are committed to providing and maintaining a safe and healthy working environment for our employees, visitors, and all people using our premises as a workplace.

To ensure a safe and healthy work environment, we will develop and maintain a health and safety management system. Specifically, management will:

1. Set health and safety objectives and performance criteria for all managers and work areas
2. Annually review health and safety objectives and managers' performance against these
3. Actively encourage the accurate and timely reporting and recording of all incidents and injuries
4. Investigate all reported incidents and injuries to ensure all contributing factors are identified and, where appropriate, plans are developed to take corrective action
5. Actively encourage people to report any pain or discomfort early on
6. Provide a treatment and rehabilitation plan that ensures a safe, early and durable return to work

7. Identify all existing and new hazards and take all practicable steps to eliminate, isolate or minimize the exposure to significant hazards
8. Ensure all employees are aware of the hazards in their work area and are adequately trained to enable them to perform their duties in a safe manner
9. Encourage employee consultation and participation in all matters relating to health and safety
10. Promote a system of continuous improvement – this includes reviewing policies and procedures each year
11. Meet our obligations under the Health and Safety in Employment Act 1992, the Health and Safety in Employment Regulations 1995, codes of practice, and any relevant standards or guidelines.
12. Every manager, supervisor or foreperson has a responsibility for the health and safety of employees working under their direction.

13. Every employee is expected to share in this commitment to health and safety in the workplace by:
14. Observing all safe work procedures, rules and instructions
15. Reporting any pain or discomfort early on
16. Taking an active role in the company's treatment and rehabilitation plan, to ensure an "early and durable return to work"
17. Ensuring all incidents, injuries and hazards are reported to the appropriate person.
18. The health and safety committee includes senior management representatives and union and other nominated employee representatives. It is responsible for implementing, monitoring, reviewing and planning health and safety policies, systems and practices

## Acronyms Health and Safety Policy

AART – Apply Advanced Resuscitation Techniques

AFARP – As far as reasonably practical

ALARA – As Low as Reasonably Achievable

ALARP – As Low as Reasonably Practicable

ASSE – American Society of Safety Engineers

BBS – Behavioral Based Safety

COP – Code of Practice

CBT – Competency Based Training

CIAED – Course in Automated External Defibrillation

DGHS – Dangerous Goods and Hazardous Substances

DIFR – Disabling Injury Frequency Rate

DoII – Department of Labor NZ

EHS, EHSQ – This time, adding the E in there means "Environment" and the Q for "Quality". This adds a layer of environmental considerations to workplace health and safety. When you see this then you know it's about systems rather than people

EHSR – Elected Health and Safety Representative

ELCB – Earth Leakage Circuit Breaker

EMP – Emergency Management Plan

ERT – Emergency Response Team

FAI – First Aid Incident

FIFR – Fatal Injury Frequency Rate

FIGJAM – F$%# I'm Good, Just Ask Me

HAZOP – Hazard and Operability

HFA – Hazard Factor Assessment

HIRA – Hazard Identification Risk Assessment

HSE – Health & Safety Executive UK

HSR – Health and Safety Representative

HSSE – Health, Safety, Security & Environment

ISHR – Industry Safety & Health Representative

JSA – Job Safety Analysis (risk assessment before starting work)

JSEA – as for JSA but includes Environmental risks

L2RA – Level Two Risk Assessment

LOTO – lock out tag out

LTFR – Lost Time Frequency Rate

LTI – Lost Time Injury

MSDS – Material Safety Data Sheet

MTI – Medically Treated Incident

NLTPHRW – National License to Perform High Risk Work

NMI – Near Miss Incident

NSCA – National Safety Council of Australia

NSFW – Not Safe for Work

OFA – Occupational First Aid

OHS – Occupational Health and Safety

## About the Author

Jackson W. Everhart has spent over thirteen years immersed in the world of health, safety, and environmental (HSE) preparedness. Renowned for his meticulous approach to worst-case scenario planning, Jackson has become a leading voice in the prepper community, dedicating his career to teaching individuals and organizations how to safeguard themselves against a myriad of potential threats.

Born and raised in the rugged terrains of the Pacific Northwest, Jackson's early life was steeped in a culture of self-reliance and respect for nature's power. This background laid the foundation for his deep commitment to survivalist and preparedness.

After earning his degree in Environmental Science with a focus on emergency management, he embarked on a career that has seen him tackle some of the most challenging aspects of HSE planning.

Jackson's professional journey began in the public sector, where he worked with various governmental agencies to develop comprehensive emergency response plans. His expertise quickly became apparent, and he was soon sought after by private corporations and NGOs for his ability to create robust safety protocols that addressed everything from natural disasters to industrial accidents. His role often involved conducting rigorous risk assessments, designing detailed response strategies, and training teams to execute these plans under pressure.

Throughout his career, Jackson has also been an avid educator, believing that knowledge is the cornerstone of effective preparedness.

He has conducted numerous workshops, seminars, and training sessions across the United States, sharing his insights and practical advice with thousands of individuals. His teaching methods are renowned for their clarity, practicality, and the real-world applications of the scenarios he presents.

Jackson's writing career began as an extension of his teaching. He started by contributing articles to various survivalist magazines and HSE journals, where his well-researched and accessible writing style quickly garnered a loyal readership. This led to the publication of his first book, which was met with critical acclaim for its thoroughness and practical utility.

"The Ultimate Prepper's Survival Bible" represents the culmination of Jackson's extensive experience and unwavering dedication to HSE preparedness.

In this comprehensive guide, he distills over a decade of expertise into actionable strategies and step-by-step plans designed to help readers navigate and survive any emergency. The book covers a broad spectrum of scenarios, from natural disasters like earthquakes and hurricanes to man-made crises such as chemical spills and active shooter situations.

Jackson's personal philosophy centers around the belief that preparedness is not just about surviving but thriving in the face of adversity. His holistic approach incorporates physical, mental, and emotional readiness, ensuring that individuals are equipped to handle the full spectrum of challenges that emergencies can present. He emphasizes the importance of community and collaboration, advocating for neighborhood preparedness programs and mutual aid networks as vital components of comprehensive survival strategies.

When he's not writing or teaching, Jackson can be found practicing what he preaches in his own life.

An avid outdoorsman, he enjoys backpacking, camping, and honing his survival skills in the wilderness. He is also deeply involved in his local community, volunteering with emergency response teams and contributing to local disaster preparedness initiatives.

Jackson W. Everhart's dedication to helping others prepare for the unexpected is more than a career—it's a calling. Through his books, teachings, and community efforts, he continues to inspire and empower people to take control of their safety and well-being, ensuring they are ready to face whatever challenges the future may hold.

www.ingramcontent.com/pod-product-compliance
Lightning Source LLC
Chambersburg PA
CBHW071921210526
45479CB00002B/498